# Oxford International Primary Geography

# Workbook

Terry Jennings

OXFORD
UNIVERSITY PRESS

Great Clarendon Street, Oxford, OX2 6DP, United Kingdom

Oxford University Press is a department of the University of Oxford. It furthers the University's objective of excellence in research, scholarship, and education by publishing worldwide. Oxford is a registered trade mark of Oxford University Press in the UK and in certain other countries



First published in 2015



British Library Cataloguing in Publication Data
Data available

978-019-831-011-2

30 29 28 27 26 25 24 23 22 21

Paper used in the production of this book is a natural, recyclable product made from wood grown in sustainable forests. The manufacturing process conforms to the environmental regulations of the country of origin.

Printed in Hong Kong by Sheck Wah Tong Printing Press Ltd.

**Acknowledgements**
The publishers would like to thank the following for permissions to use their photographs:

Cover photo: Getty Images/Reinhard Dirscherl

Although we have made every effort to trace and contact all copyright holders before publication this has not been possible in all cases. If notified, the publisher will rectify any errors or omissions at the earliest opportunity.

# Contents

# Maps

## A world map

**Key**

| | | |
|---|---|---|
| **1** Canada | ☐ | Orange |
| **2** United States of America | ☐ | Blue |
| **3** Brazil | ☐ | Black |
| **4** Russia | ☐ | Red |
| **5** China | ☐ | Yellow |
| **6** India | ☐ | Brown |
| **7** Australia | ☐ | Green |
| **8** Saudi Arabia | ☐ | Purple |

Here is a map of the world. It shows only some of the larger countries.

- Colour the boxes in the key.
- Colour the countries shown on the map, using the correct colours.

# Village maps

Write the names of these maps in order, starting with the smallest-scale map.

**The world**

**Frigiliana centre**

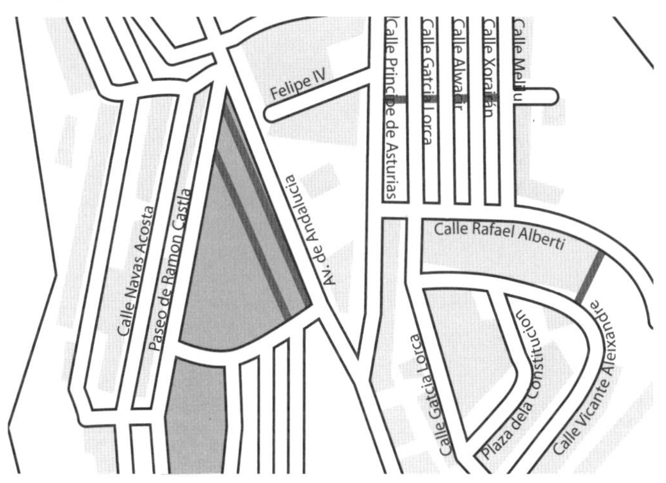


**Europe**

**Spain**

**Frigiliana**

1 ______________________________

2 ______________________________

3 ______________________________

4 ______________________________

5 ______________________________

# Directions

## Finding the way

You can use a compass to find directions.
A compass needle always points to north.

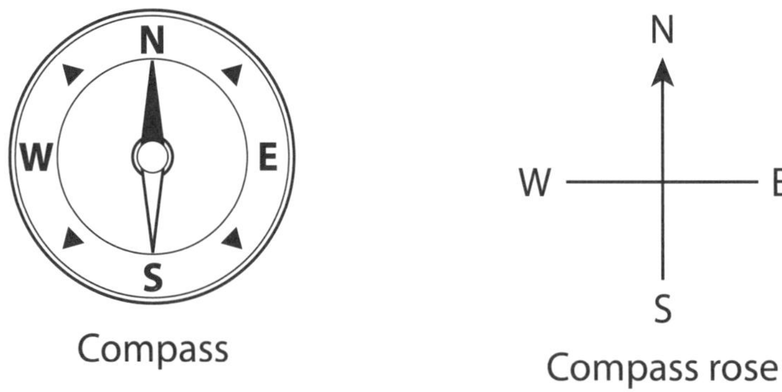


Compass

Compass rose

You can use a compass rose to find directions on a map.
A compass rose points north, south, east and west.

You will need a compass for this activity.

- Put your compass on the tower in the middle of the box. Turn your compass until the needle points towards 'N'. This is north.
- Draw an arrow pointing north next to the tower.
- To the east of the tower draw a tree.
- To the west of the tower draw a street light.
- To the north of the tower draw a flagpole and flag.
- To the south of the tower draw a bridge.

# Compass directions

Fill in the directions on this compass rose.

Use the words from the word box to help you.

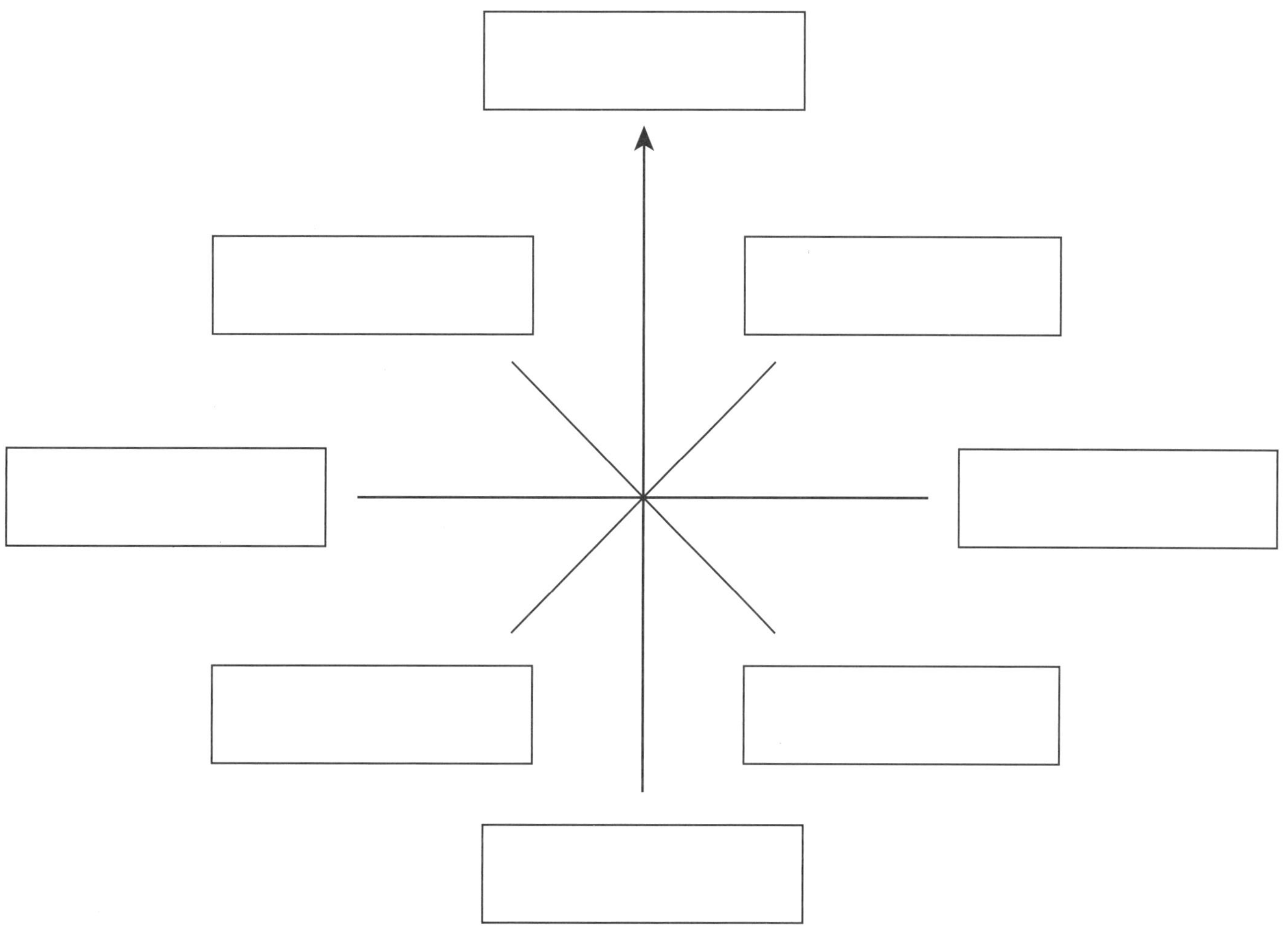

| | | | |
|---|---|---|---|
| **West** | **North-East** | **South-West** | **East** |
| **North-West** | **South** | **North** | **South-East** |

1 Which direction does the front of your school face?

______________________________

2 In which direction is your home from your school?

______________________________

# Villages

## Village, town or city?

Write the words from the word box in the correct column in the table. Some words will go in both columns.

**market**
**town hall**
**cinema**
**embassy**
**post office**
**children's play area**
**parliament building**
**hospital**
**university**
**railway station**
**village hall**
**garage**
**park**
**bank**
**museum**
**school**
**coffee shop**
**farm**
**apartments**
**superstore**
**trees**
**bus station**
**palace**
**office block**
**health centre**
**factory**

| Village | Town or city |
|---|---|
| | |

Add some words of your own if you can.

# Plan of a village

Work with a friend to design a village on the map below.

Copy the symbols from the key.

You can use each symbol as many times as you want.

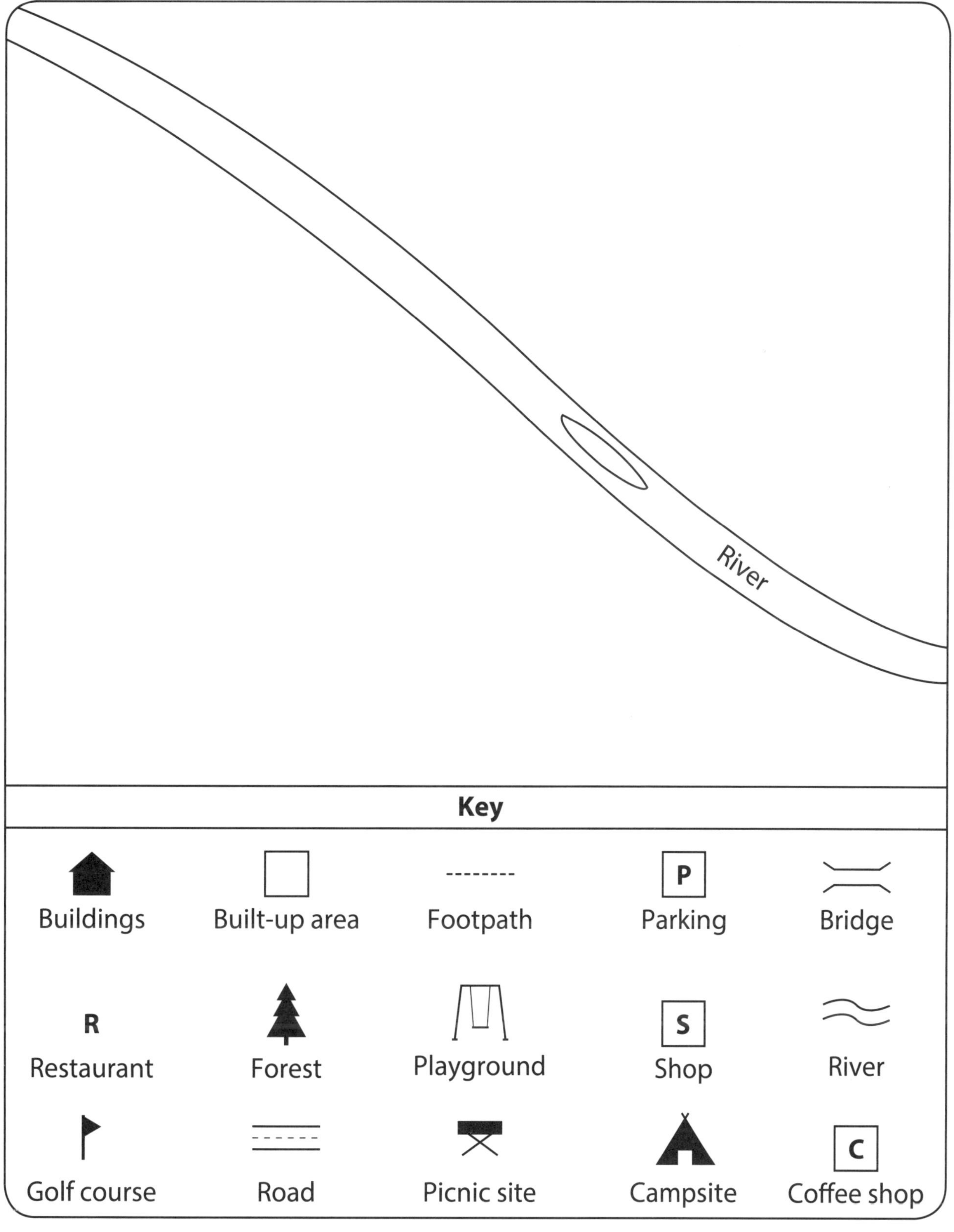

# The village of Dana

## Dana, a village in Jordan

Here is a map showing the location of the village of Dana in Jordan.

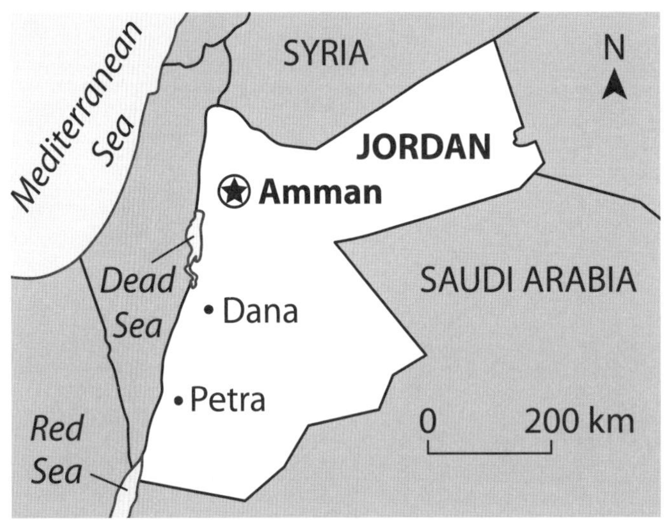


Answer these questions:

**a** Is this a large-scale map or a small-scale map? ____________________

**b** How do you know? ____________________

**c** Name two seas that form part of Jordan's borders with other countries ____________________

____________________

**d** Which is the smallest of the three seas shown on the map? ____________________

____________________

**e** In which direction is the capital city, Amman, from Dana? ____________________

____________________

**f** Roughly how far is it from Amman to Dana? ____________________

**g** In which direction is Petra from Dana? ____________________

**h** In which direction would you have to travel from Dana to reach Syria? ____________________

____________________

# Jordan fact file

Dana is a village in Jordan. Find out all you can about Jordan.

Fill in the fact file.

| Continent | |
|---|---|
| Population | |
| Capital city | |
| Currency (money) | |
| Flag | |
| Languages spoken | |
| Other big cities | |
| Famous sights and interesting facts | |

# Farms and food

## Food from the farm

Draw lines to match the pictures of what is grown on a farm to what we buy in a shop. Some items from the first column will link to more than one item in the second column.

On the farm

In a shop

# Two different farms

Look carefully at the pictures.

Decide what type of farm you can see.

Write a description of each farm and colour the pictures.

France

**Type of farm**

**Type of farm**

China

# Choosing a holiday

## Holiday countries

Choose five countries that people go to on holiday.

For each country, name the capital city and the currency (money), and draw the flag.

Use an atlas, reference books or the Internet to help you.

| Country | Capital city | Currency | Flag |
|---|---|---|---|
| | | | |
| | | | |
| | | | |
| | | | |
| | | | |

# My dream holiday

Imagine you could go anywhere you wish for a holiday. Which country would you go to?

Do some research using holiday brochures and the Internet and then write out your holiday plans. The country I would like to go to is ____________________

The month I would go in is ____________________

The climate during this month is ____________________

The people who live in this country are called ____________________

The currency (money) I would need there is ____________________

The cities I would like to visit are ____________________

The sights I would like to see are ____________________

One food I would like try is ____________________

The activities I would like to do are ____________________

The reason I chose this country is because ____________________

Stick a map of the country below. Mark on it the cities you would like to visit.

# Hot and cold places

## Thermometers and weather

Look at the picture. It shows a thermometer. We use a thermometer to tell us how hot or cold something is. The thermometer in the picture measures temperature in degrees Celsius. We write this as °C. This type of thermometer is used to measure the temperature of the air. What temperature does this thermometer show? ______________________________

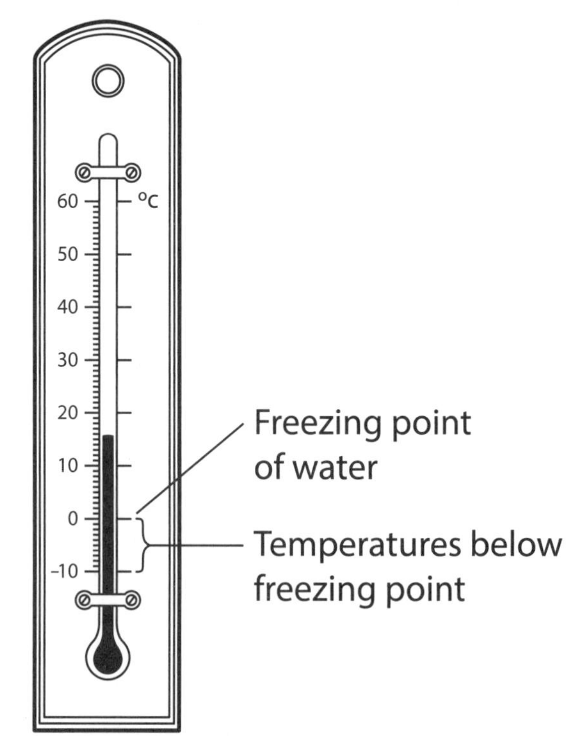


Here are pictures of four places.

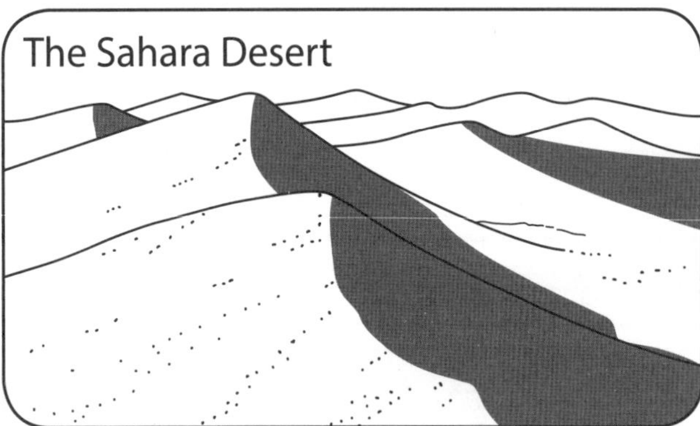

Use reference books, atlases, holiday brochures or the Internet to find out what the weather is like in these places in June and December.

# Differing climates

The climate changes according to where on the Earth you live.

a

Where I live is cold all the year round. There is ice and snow everywhere.

b

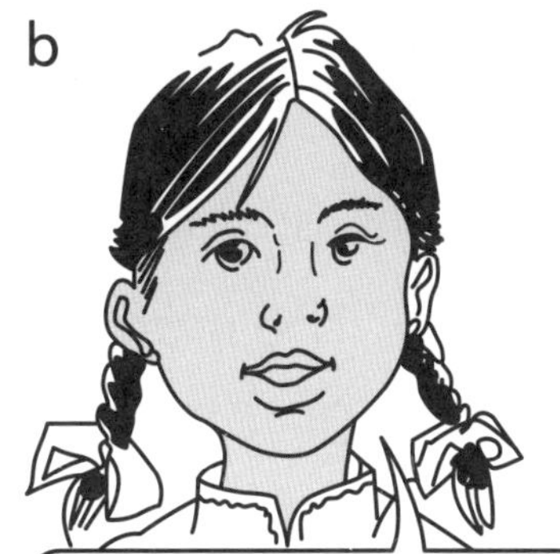

It is hot where I live. At certain times of the year there are heavy rains.

c

Where I live it is hot, sunny and dry all the year round.

d

Where I live it is not very hot or very cold. The weather is changeable.

Match the children above to the countries where they live.
Write the letter in the box

☐ India ☐ Greenland ☐ France ☐ Egypt

You are going to visit one of the four countries above. Why might you need to know what the weather is going to be like? Write a list of the reasons.

________________________________________

________________________________________

________________________________________

________________________________________

________________________________________

# Looking at Switzerland

## Putting Sisikon on the map

Sisikon is a village in Switzerland. It is on the shores of Lake Lucerne.

Number the maps 1 to 5. The map with the smallest scale should be number 1 and the map with the largest scale should be number 5.

**The world**

**Europe**

**Sisikon**

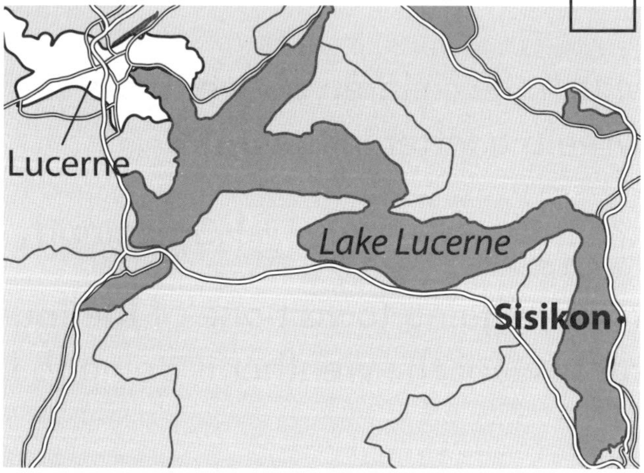

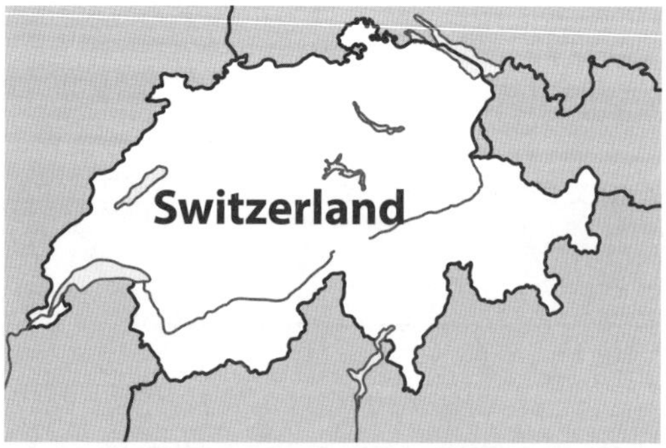

# Switzerland fact file

Find out all you can about Switzerland.

Fill in the fact file.

| Continent | |
| --- | --- |
| Population | |
| Capital city | |
| Currency (money) | |
| Flag | |
| Languages spoken | |
| Other big cities | |
| Famous sights and interesting facts | |

# Life in Cairo

## Cairo and Egypt

Here is a map of Egypt.

1 Use an atlas to help you add details to the map.

- Label Libya and Sudan.
- Label the capital city of Egypt (shown as ⊛ on the map).
- Write in the name of the river.

2 Does this river flow towards the north or the south? How do you know?

________________________________________

3 Name the two seas that are shown on the map.

________________________________________

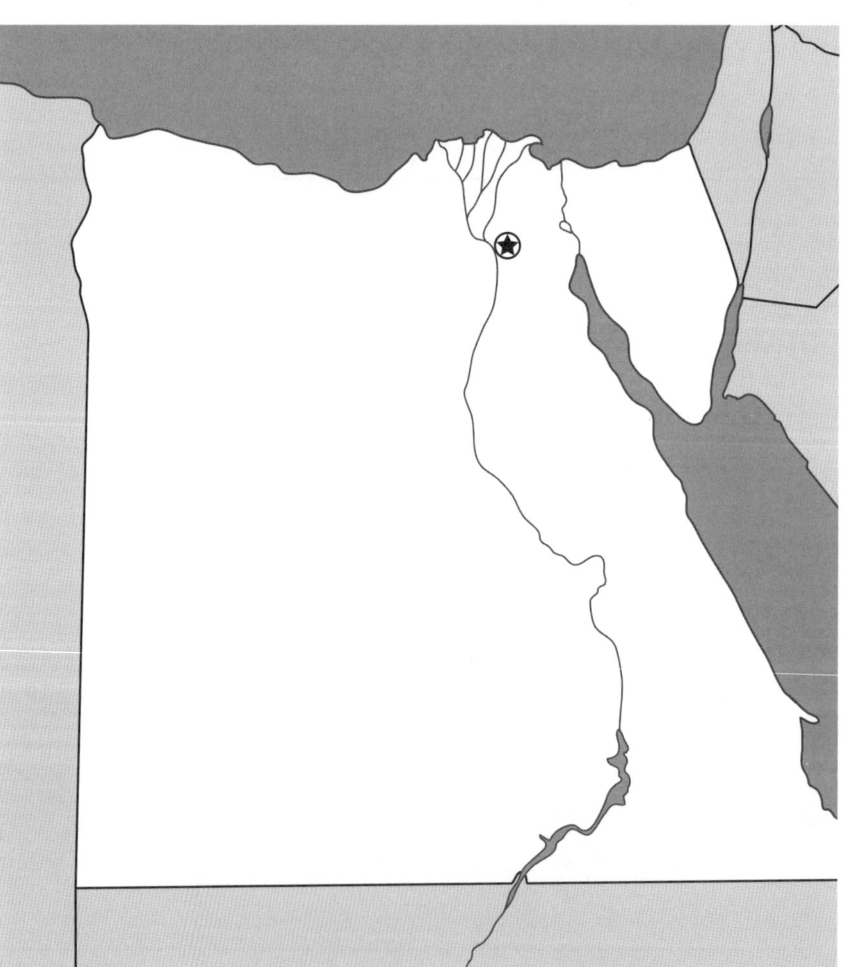

# Egypt fact file

Find out all you can about Egypt.

Fill in the fact file.

| Continent | |
|---|---|
| Population | |
| Capital city | |
| Currency (money) | |
| Flag | |
| Languages spoken | |
| Other big cities | |
| Famous sights and interesting facts | |

## A local home

Have you ever looked really closely at your own home? Go outside and study it carefully, then complete this record sheet.

- Name of street or road your home is in: ______
- Number or name of your home: ______
- Number of floors: ______
- What type of building is it (apartment, villa or house, terraced house, cottage, bungalow, caravan)? ______
- What building materials are used for:

the roof? ______ the walls? ______

the doors? ______ the window frames? ______

the drainpipes? ______

- Does it have:

a chimney? ______ a garage? ______ a porch? ______

a driveway? ______ a garden? ______ a front gate? ______

Write down any other features that you can see (for example a television aerial, a satellite dish, window boxes, a date plaque).

______

______

- Now draw a picture of your home.

# People who work in my street

Think about all the people who work in, or visit, your street.

For each person, or group of people, write down who relies on them and why.

| Workers in my street | Who relies on them? | Why? |
|---|---|---|
| *Road sweeper* | *Everyone* | *To keep the streets clean* |
| | | |
| | | |
| | | |
| | | |
| | | |
| | | |
| | | |
| | | |
| | | |

Which of the people above help to improve your street?
Tick (✓) their names.

## Using land

Land is used in many ways. Farms use large areas of land. Small shops cover only small areas of land.

Look at the pictures on this page.

Decide whether each one uses a small or a large area of land.

Complete the two lists. They have already been started for you.

| Small area of land | Large area of land |
|---|---|
| Baker's shop | Farm |

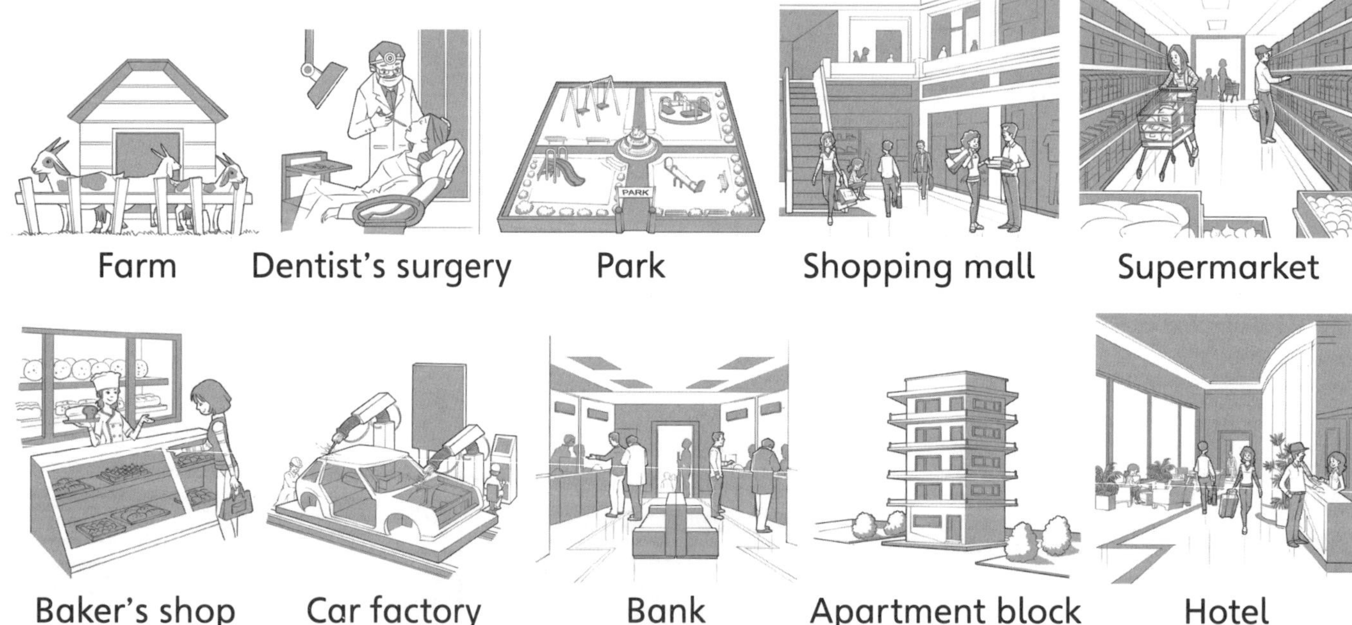


Farm Dentist's surgery Park Shopping mall Supermarket

Baker's shop Car factory Bank Apartment block Hotel



Compare your answers with those of your friends.

Add your own ideas to the lists.

# Local land uses

Some children went to find out what the land was used for in a local village. The map shows what they found.

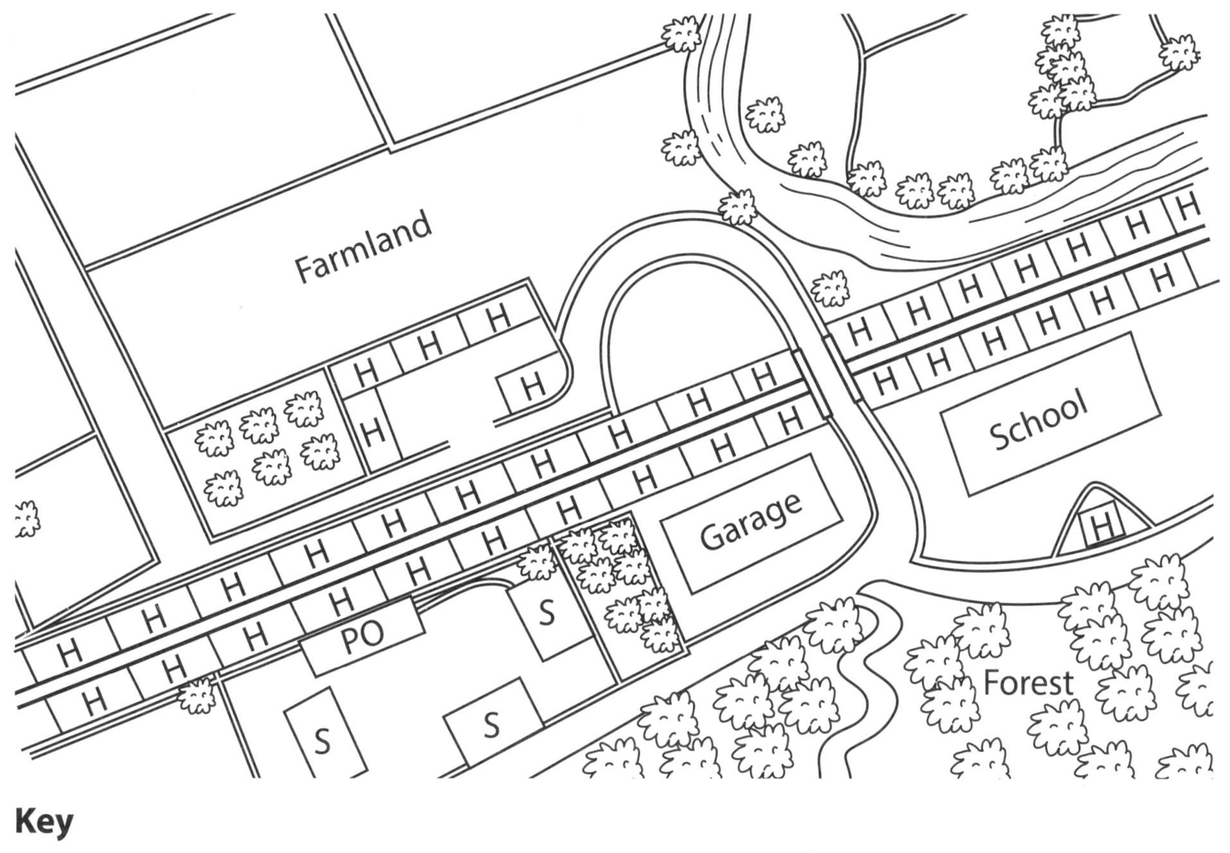


**Key**

S = Shops PO = Post Office H = House Tree

**1** Colour in the different land uses on the map. Use the colours shown in the key below. Colour each box in the key with the matching colour.

**COLOUR KEY**

| | | |
|---|---|---|
| ☐ Red | for houses | ☐ Green for forest |
| ☐ Yellow | for services (buildings that people use, but not houses and not shops) | ☐ Brown for farmland |
| ☐ Blue | for shops | ☐ Black for roads |

**2** Look at the map again, and then answer these questions.

How many houses are there? ____________________

How many services are there? ____________________

How many roads are there? ____________________

# People and the environment

## The school grounds

Look carefully around your playground.

How many litter bins are there? ______________________

Who empties them? ______________________

How often are they emptied? ______________________

How often is the rubbish collected from your school? ______________________

What litter can you find on the playground? (Tick the boxes.)

- [ ] Sweet wrappers
- [ ] Lollipop sticks
- [ ] Apple cores
- [ ] Orange peel
- [ ] Banana skins
- [ ] Plastic bottles
- [ ] Crisp bags
- [ ] Scraps of paper
- [ ] Empty cans

Other (list the items) ______________________

______________________

______________________

Each year an average school throws away approximately 45 kilograms of rubbish per pupil – that's a lot of rubbish!

What can you do to make sure your school throws away less rubbish? ______________________

______________________

______________________

# Animals at risk

The animals below are all considered to be endangered.

This means there is a risk that they could become extinct.

Use the Internet and reference books to find out why the numbers of these animals are falling.

Suggest things that can be done to help save these animals from becoming extinct.

| Animal at risk | Reasons why it is at risk | Possible ways of saving it |
|---|---|---|
| Blue whale | | |
| African elephant | | |
| Black rhinoceros | | |
| Scarlet macaw | | |

# The changing seasons

## The seasons

Look at the picture below. It shows heat and light from the Sun reaching the Earth.

The Earth's axis, around which it turns each day, is tilted. In the picture the positions of France and Australia are marked on the Earth.

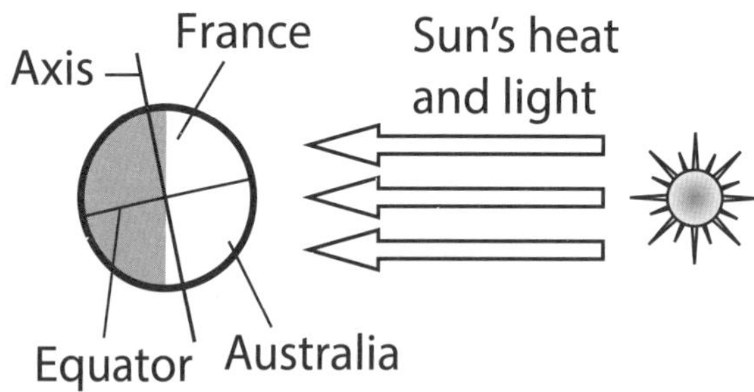


1 a Does the picture show summer or winter in France? ______________

b How do you know? ______________

2 a Does the picture show summer or winter in Australia? ______________

b How do you know? ______________

3 Draw a picture like the one above to show how the Earth's axis is tilted when it is summer in France.

4 Imagine that the Earth's axis was not tilted (as in the picture below).

What would the seasons be like in:

a France? ______________

b Australia? ______________

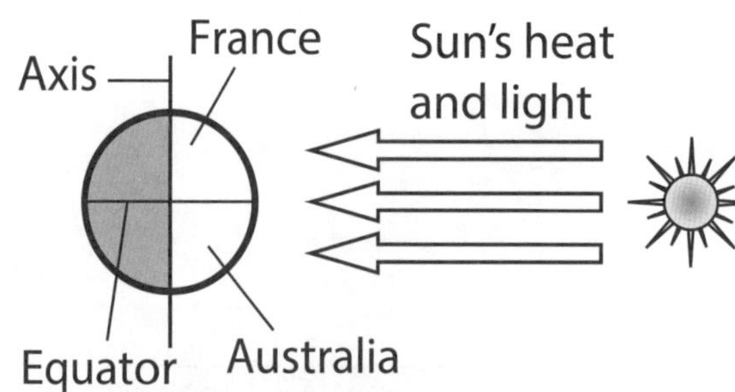

# The changing seasons

Many parts of the world have four seasons each year: spring, summer, autumn and winter.

Choose one season. Write how your season affects the environment – people, transport, animals, plants and land.

I have chosen ____________________

This is how it affects the environment.

People ____________________

____________________

____________________

Transport ____________________

____________________

____________________

Animals ____________________

____________________

____________________

Plants ____________________

____________________

____________________

Land ____________________

____________________

____________________

# Keeping in touch

## Addressing a letter

1 Address one of the envelopes to yourself. Write:
   - your name
   - your house number or name and street name
   - the city, town or village
   - the area or state
   - the postcode or zip code.

2 How would you address a letter to your teacher if you were sending it from the Moon?

3 Write this address on the other envelope.

# The local post box

Visit your local post box or mail box with a responsible adult.

It is where you put letters that you want to be sent to other people.

Where is the post box you have chosen? ______________________

Is it inside a building or out in the open? ______________________

Is it built into a wall, fixed to a wall or standing on its own? ______________

What colour is it? ______________

Draw your post box.

Look carefully at the front of the post box.

Does it tell you when the box is emptied?

______________________

Write down the collection times.

Collection times

At what time is the first collection each day? ______________________

How many collections are there each weekday? ______________________

How many collections are there at the weekend? ______________________

# Electronic mail (e-mail)

## Sending and receiving messages

A letter is a way of sending a message. Here are some other ways. What are they?

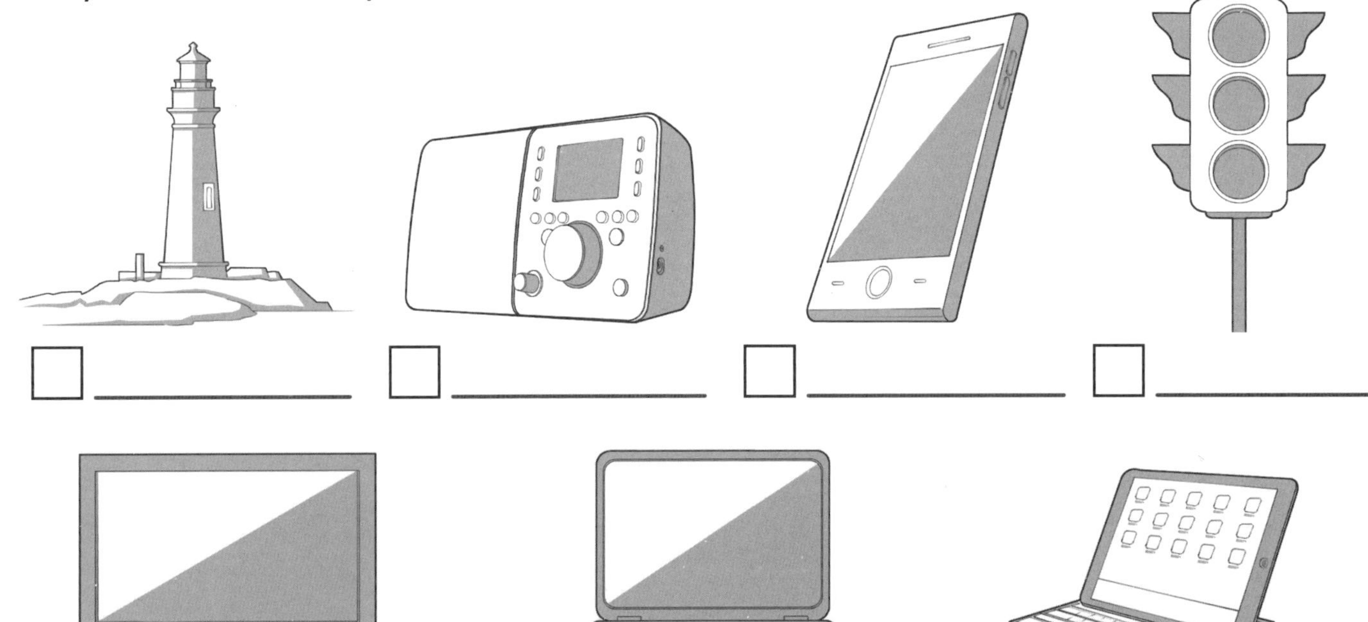

1 For each of the ways of sending a message, write in the boxes whether you would:

a see the message **(S)**

b hear the message **(H)**

c both see and hear the message **(B)**.

2 Write a text message to a friend in a school in another part of the country. Include three questions you would like to ask that person about the place where he or she lives.

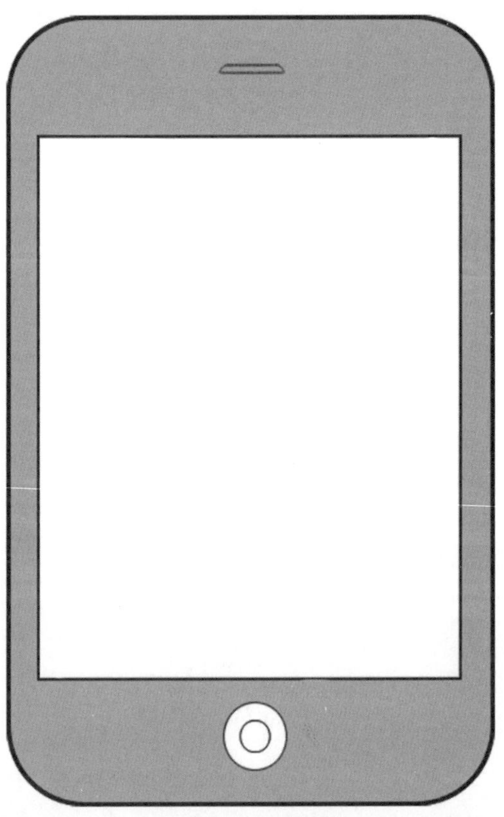

# E-mail a friend in another country

Imagine you have been sent this e-mail from two students in another country.

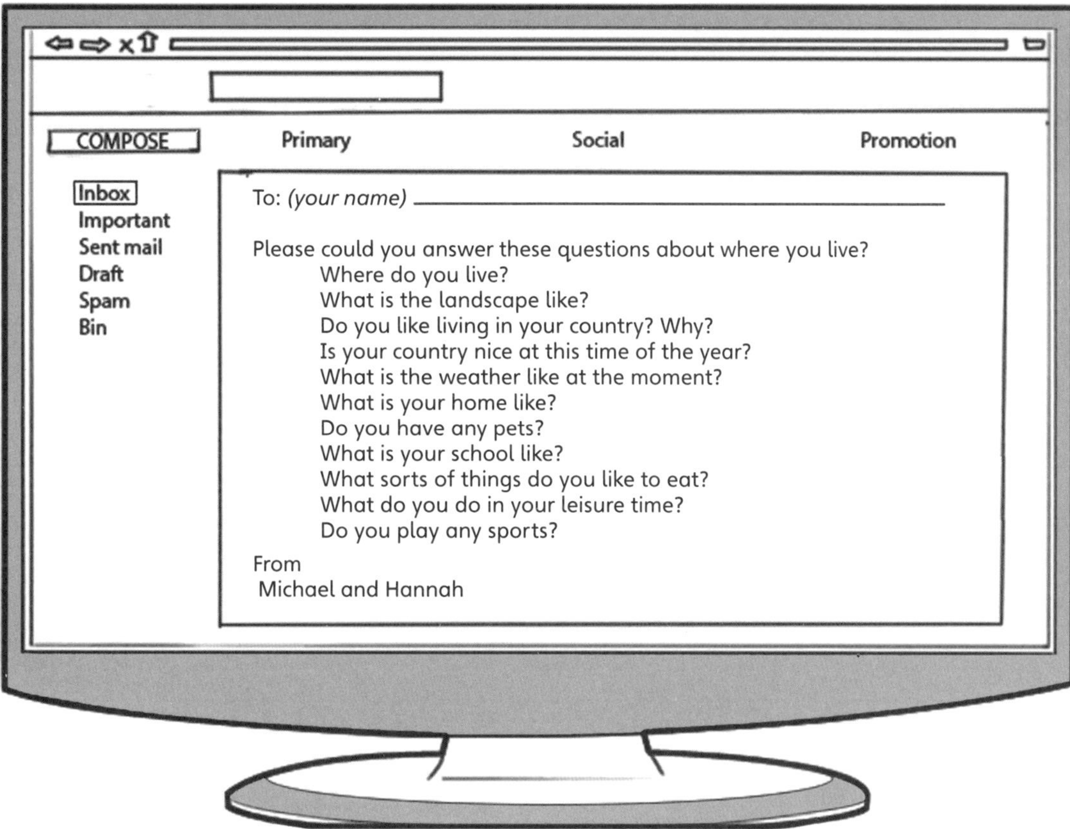


Write your reply as an e-mail.

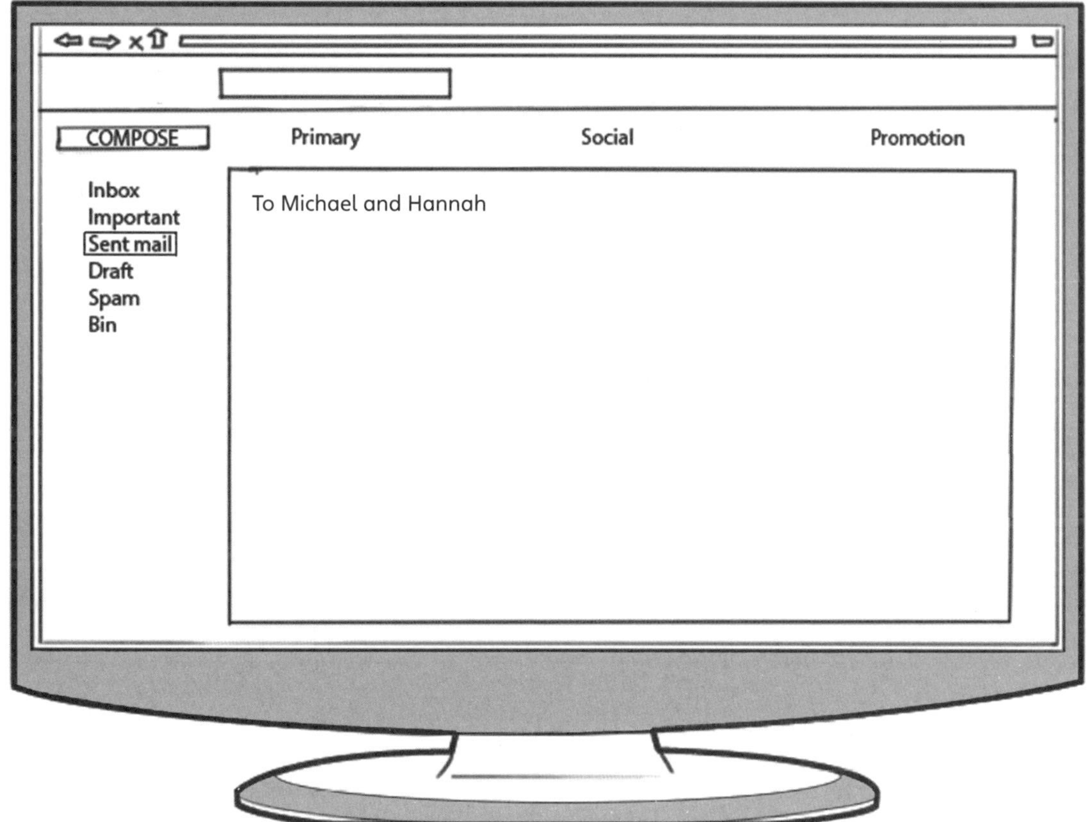

# Daily journeys

## My journey to school

Ask the students in your class how they came to school. Write the number of students who used each of these different ways of travelling.

| Way of getting to school | Number of students |
|---|---|
| Walking | |
| In a car or taxi | |
| Using the train or Metro | |
| On a bus | |
| On a tram | |
| On a bicycle | |

Draw a bar chart to display your results.

Use your bar chart to answer these questions:

1 What is the most popular way of travelling to school? ____________

2 What is the least popular way of travelling to school? ____________

3 Write one advantage and one disadvantage of each way of travelling to school.

| Way of getting to school | Advantage | Disadvantage |
|---|---|---|
| Walking | | |
| In a car or taxi | | |
| Using the train or Metro | | |
| On a bus | | |
| On a tram | | |
| On a bicycle | | |



Is your class typical of the whole school?

Do a survey of the whole school to find out.

# Traffic and the environment

Carry out a survey of the traffic near your school.

Copy this chart into a notebook or onto a piece of paper (or ask your teacher to photocopy it onto a piece of paper).
Each time you see one of the vehicles in the list put a tick in the number column.

Date: ____________ Time: ____________ Location: ____________

| Type of vehicle | Number | Total |
|---|---|---|
| Bicycle | | |
| Motor cycle or scooter | | |
| Car carrying one person | | |
| Car carrying two people | | |
| Car carrying more than two people | | |
| Van | | |
| Truck | | |
| Bus | | |

Use your survey to answer these questions.

Which type of vehicle is the most common? ____________

Which type(s) of vehicle are bad for our environment? Say why. ____________

____________

Which type(s) of transport are best for our environment? Say why. ____________

____________

If people have to use cars, suggest ways of using them that would be better for the environment. ____________

____________

# Weather in the news

## Extremes of weather

- Watch the television news for stories about the weather. Watch the weather reports. Study the newspapers and the Internet for news about the weather.
- Record any extremes of weather, such as floods, droughts, storms, blizzards, hurricanes and tornadoes. Fill in a box for each one saying what happened.
- Draw a line to each country on the map of the world to show where it happened.

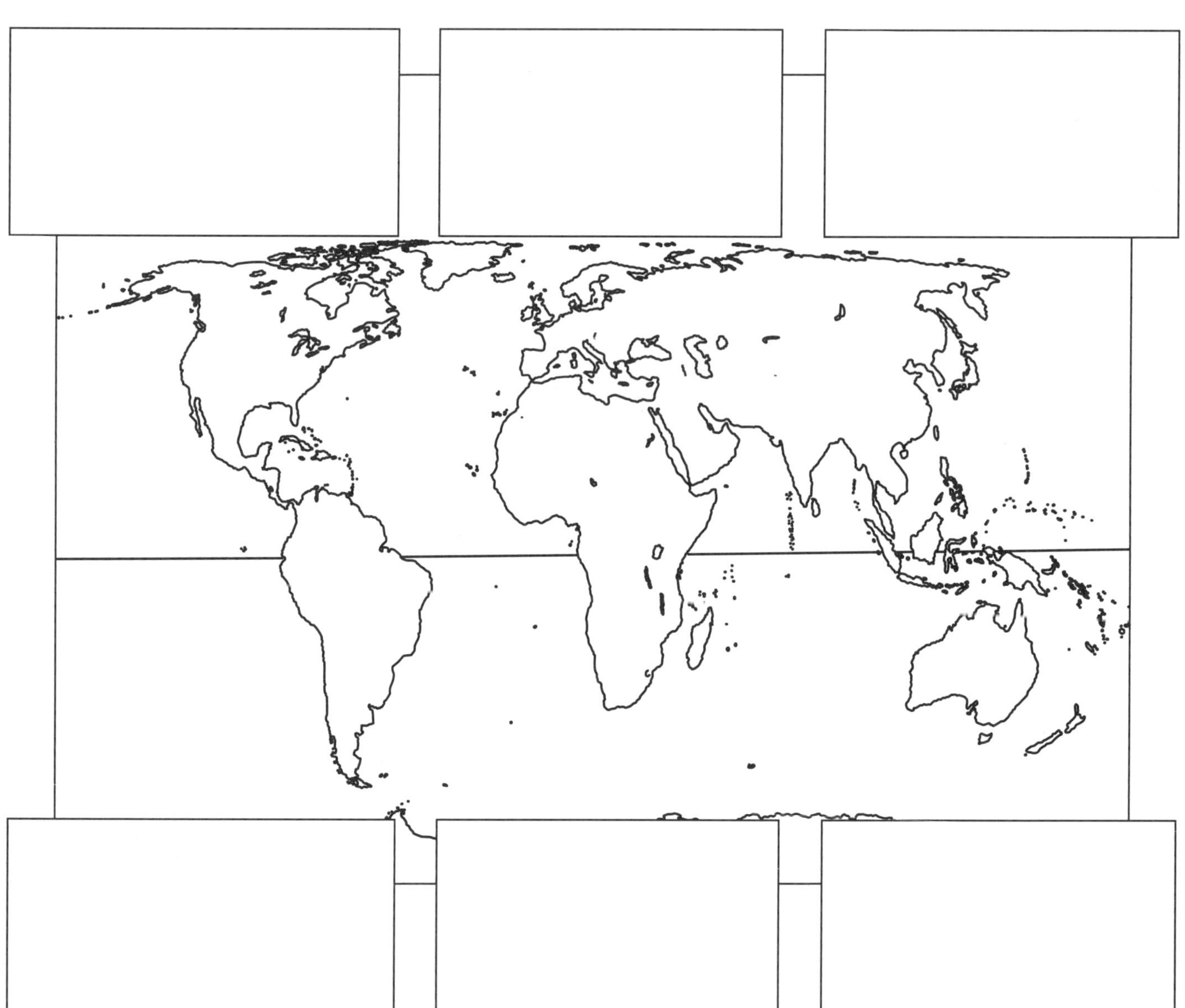

# Flood!

Imagine that you live in a house by a river. One day the river floods. You will be safe upstairs, but you may have to stay there for several days. You just have time to collect ten things from inside the house to help you survive.

Write a list of the things you will collect. Against each one, say why you have chosen it.

In a separate book, write the story of the flood. Describe what happens to you and your house.

Draw a picture of your house in the flood.

# Changing weather conditions

## It's raining!

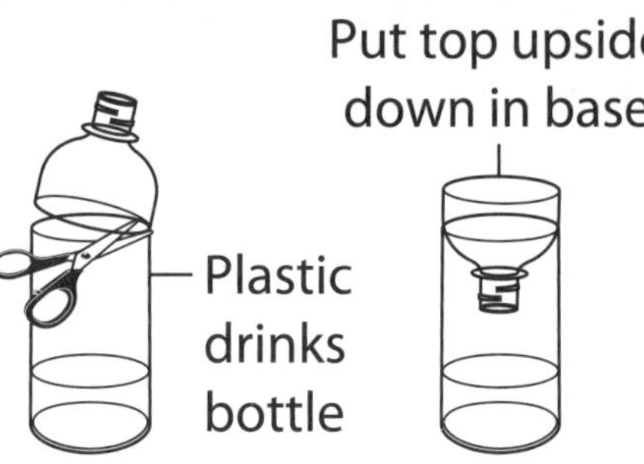


Work in a group to make four rain gauges like this one.

- Place the rain gauges outside in four different places.
- Measure how much rainwater the gauges contain at the end of each week, for three weeks. To do this, tip the water from each gauge into a measuring cylinder marked with millilitres (ml) and measure the amount of the water.
- Record your results in this table.

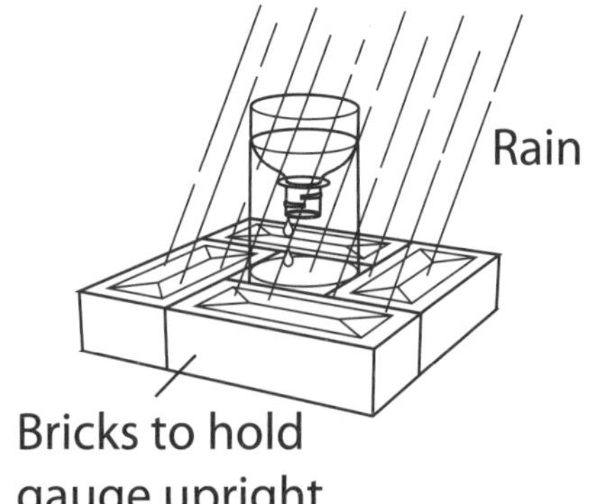


| Location of rain gauge | Week 1 | Week 2 | Week 3 | Total rainfall (ml) |
|---|---|---|---|---|
| 1 | | | | |
| 2 | | | | |
| 3 | | | | |
| 4 | | | | |

- Draw a bar chart to show the total rainfall that was collected by each rain gauge in three weeks.
- Use the bar chart to help you to answer these questions:

1 Which rain gauge had collected the most water at the end of three weeks?

______________________________

2 Which rain gauge had collected the least amount of rain after three weeks?

______________________________

3 Which week had the most rainfall? ______________________________

4 Did the position of your rain gauges affect the amount of rainwater collected? Say how. ______________________________

# A weather chart

Complete the weather chart every morning and afternoon for a week using these symbols.

sun

thunderstorm

rain

cloud

snow

wind

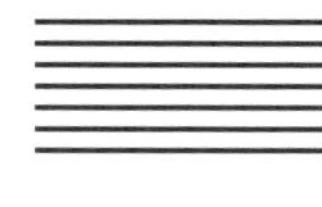
fog

At the end of each day, fill in the last column.

This is your weather forecast for the next day.

| Day | Morning | Afternoon | Temperature | Wind direction | Rain | Tomorrow the weather will be: |
|---|---|---|---|---|---|---|
| Monday | | | | | | |
| Tuesday | | | | | | |
| Wednesday | | | | | | |
| Thursday | | | | | | |
| Friday | | | | | | |

# Local changes

## Planning permission

In most countries, people must have planning permission from the local council or government before they can construct a new building or change an existing building.

In areas where there are interesting old buildings, planning officers control any changes to the buildings and streets.

Here are some of the rules. Think about them carefully. Are they good or bad ideas? For each one, tick whether you think it is a good or bad idea and write why you think this.

| Rule | Good idea | Bad idea | Why? |
|---|---|---|---|
| Brickwork and stonework must be kept the same colour as they were originally. | | | |
| New extensions must be in the same style as the original building. | | | |
| New windows and doors must be the same as the original ones. | | | |
| New roof materials must be the same as the old. | | | |
| Big trees must not be cut down to give more light or space. | | | |
| All television aerials must be in the roof. | | | |
| No satellite dishes should be visible. | | | |
| Dustbins must not be kept at the front of the houses. | | | |

## Good or bad planning?

Places that have good weather attract many tourists during the summer. Thousands of people visit these places on holiday and spend money in hotels, in restaurants, taking trips and buying souvenirs. Many local people make their living from the tourists.

In many holiday resorts, new hotels and villas are being built to cope with the growing number of tourists. This means that rich soil, which is ideal for growing crops, is being built on.

Imagine that you are **one** of the following people who live in a holiday resort. Tick the box of the person you have chosen.

- a shopkeeper ☐
- a builder ☐
- a farmer ☐
- a retired person ☐
- an environmentalist ☐
- a restaurant owner. ☐

Are you **for** or **against** the plans for new hotels and villas? List your reasons.

| FOR | AGAINST |
| --- | --- |
| | |

## Local traffic problems

What is the traffic like near your school? What do people think about it?

Choose two people, who you know, to talk to about the traffic near your school. Write their answers on this page.

1 Where do you think the most dangerous part of this road is, and why?

| Person A | Person B |
|---|---|
| | |

2 What should be done to reduce the number of vehicles or to slow them down?

| Person A | Person B |
|---|---|
| | |

3 If the road was closed to traffic, what do you think would be the advantages? What would be the disadvantages?

| Person A | Person B |
|---|---|
| | |

# Traffic and the weather

Here are eight different kinds of transport.

Which kinds of weather can delay each different kind of transport?

Fill in the chart below. You can use some or all of the words from the word box.

| | | | |
|---|---|---|---|
| **ice** | **snow** | **fog** | **mist** |
| **rain** | **dust storm** | **gale** | **thunderstorm** |
| **hurricane** | **tornado** | **hail** | **sleet** |

| Car | Truck | Motor cycle | Bicycle | Train | Ship | Bus | Aircraft |
|---|---|---|---|---|---|---|---|
| | | | | | | | |

Which kinds of transport are most easily delayed because of the weather?

______________________________________________

______________________________________________

# Map of the World

# Glossary

**Aerial photograph** A photograph taken from above, usually from an airplane, balloon or a drone.

**Atlas** A book of maps.

**Capital** The most important city in a country.

**Climate** The typical weather of a place over the whole year.

**Commuter** Someone who regularly travels to work, especially by bus or train.

**Compass** A device for showing which direction is which; the needle always points north.

**Crop** A plant grown by farmers.

**Electronic mail (e-mail)** The use of a computer system to send or receive messages over long distances.

**Endangered** An animal or plant that is in danger of dying out.

**Environment** Your surroundings.

**Equator** An imaginary line round the Earth at an equal distance from the North and South Poles.

**Extinct** Not existing anymore.

**Farm** The land and buildings used for growing crops and looking after animals.

**Fertile** Able to produce good crops.

**Fertiliser** A chemical put into the soil to make crops grow better.

**Fuel** Anything that can be burned to produce heat or light.

**Globe** A ball with a map of the whole world on it.

**Gorge** A narrow, deep valley with steep sides.

**Hay** Dried grass used for feeding farm animals.

**Irrigation** To put water onto the land so that crops grow well.

**Key** The key to a map tells you what the various symbols on it mean.

| | |
|---|---|
| **Meadow** | An area of natural grassland, often used for grazing animals. |
| **Meteorologist** | A scientist who studies the weather. |
| **Minerals** | Useful chemical substances in the ground obtained by mining or quarrying. |
| **Monsoon** | A strong wind in and around the Indian Ocean that brings heavy rain in summer. |
| **Mountain** | A very high part of the Earth's surface. |
| **North Pole** | The most northerly part of the Earth. |
| **Oasis** | A place where water can be found in the desert. |
| **Pesticide** | A chemical sprayed onto crops to stop insects and other pests from destroying them. |
| **Pollute/ Pollution** | When substances such as air, water or the soil are spoiled or made dirty by people. |
| **Postal service** | The service that carries letters and parcels. |
| **Scale** | The scale on a map tells us how much the distance on paper (e.g. 1 centimetre) is actually worth on the ground (e.g. 50 kilometres). |
| **Season** | A time of the year when you can expect a special type of weather. |
| **Settlement** | A place where people live. Villages, towns and cities are settlements. |
| **Source** | The place where a river starts. |
| **South Pole** | The most southerly part of the Earth. |
| **Spring** | A place where water naturally flows out of the ground. |
| **Symbol** | A small, simple drawing used on a map to show where things such as car parks and castles are. |
| **Temperature** | How hot or cold someone or something is. |
| **Text message** | A short message sent using a mobile phone. |
| **Tourist** | Someone visiting a place for pleasure rather than for work. |

**Traffic** Cars, buses, trucks, bicycles, and so on, travelling along a road.

**Typhoon** A huge storm that sweeps into eastern Asia from the Pacific Ocean.

**Valley** A line of low land between hills or mountains.

**Weather** The rain, wind, snow, sunshine and temperature at a particular time or place.

**Weather forecast** A prediction of what the weather is going to be like.